AF849208

Contents

1 Introduction to Smart Factories

Smart factories represent a paradigm shift in manufacturing, moving beyond traditional methods to embrace interconnected systems and intelligent automation. This isn't simply about adding robots; it's a fundamental change in how we design, manage, and optimize the entire production process. Imagine a factory floor where machines communicate seamlessly, predicting maintenance needs and adjusting production in real-time based on demand. This is the core promise of a smart factory.

The transition to a smart factory requires a careful and strategic approach. It's not a simple upgrade but a holistic transformation. Begin by assessing your current operations, identifying bottlenecks and inefficiencies. This thorough analysis will highlight areas where automation can provide the most significant improvements. Don't jump into expensive implementations without a clear understanding of your specific needs and return on investment. A phased approach, starting with pilot projects in specific areas, is generally recommended, allowing you to learn and adapt as you go.

Data is the lifeblood of a smart factory. Sensors embedded in machines and throughout the production process collect vast amounts of information, providing real-time insights into operations. This data is crucial not only for monitoring performance but also for predictive maintenance and optimizing processes. Investing in robust data analytics capabilities is essential to extract meaningful insights from this data deluge, enabling data-driven decision-making. Remember to carefully consider data security; robust cybersecurity measures are essential to protect sensitive information and prevent disruptions.

Implementing smart factory technologies requires more than just acquiring new equipment. It necessitates a cultural shift within the organization. Your workforce needs to be trained and empowered to work alongside these new technologies. This involves not only technical training on using the new systems but also fostering a culture of continuous improvement and collaboration. Employees must understand the benefits of the changes and how their roles will evolve to effectively leverage the new capabilities of the smart factory.

Integrating smart factory technologies with existing systems presents another significant challenge. Many companies have legacy systems that might not be compatible with new technologies. Carefully consider how to integrate these systems effectively without incurring significant disruptions to production. This often requires careful planning, potentially involving phased rollouts or the development of custom interfaces to bridge the gap between old and new systems. Proper planning and a clearly defined integration strategy are absolutely crucial for a successful transition.

The benefits extend beyond increased efficiency and reduced costs. Smart factories can enhance product quality through better process control and real-time monitoring. They can also improve responsiveness to market demands by enabling quicker adjustments to production based on real-time sales data. The enhanced flexibility of smart factories also allows for easier customization and personalization of products, catering to evolving customer needs. A robust supply chain integration further enhances responsiveness and minimizes disruptions caused by unforeseen events.

Consider the potential challenges. The upfront investment in new technologies and infrastructure can be substantial. It is imperative to carefully evaluate the costs and benefits before proceeding. The complexity of integrating various systems requires specialized expertise. Finding and retaining skilled workers who can operate and maintain these advanced systems is also a key concern. These factors must be carefully considered during planning, ensuring that your chosen strategies directly mitigate those potential hurdles.

Successfully implementing a smart factory requires a comprehensive approach that goes beyond simply acquiring new technologies. It demands a holistic strategy involving careful planning, comprehensive training, and a commitment to continuous improvement. Remember to assess your current capabilities and prioritize areas where automation will provide the most significant benefits. Begin with pilot projects, gradually expanding as you gain experience and confidence. Finally, always prioritize the safety and training of your workforce, embracing a culture of collaboration to reap the full potential of a smart factory environment. By adopting this pragmatic and holistic approach, you can successfully navigate the challenges and unlock the transformative potential of smart factory technology. Embrace the change, and you'll position your organization for growth and success in the increasingly competitive global marketplace.

1.1 Defining Smart Manufacturing

Smart manufacturing isn't just about adding robots; it's a fundamental shift in how we design, produce, and deliver goods. It leverages advanced technologies to create a flexible, responsive, and efficient production environment. Think of it as a dynamic ecosystem where data flows seamlessly, machines communicate intelligently, and human expertise is amplified, not replaced. This interconnectedness allows for real-time optimization, predictive maintenance, and improved decision-making across the entire value chain. Ultimately, it's about maximizing productivity while minimizing waste, improving product quality, and enhancing overall competitiveness in today's rapidly evolving market. It's a holistic approach that necessitates integrating various advanced technologies and methodologies for a comprehensive transformation.

The core of smart manufacturing lies in its data-driven nature. Unlike traditional manufacturing, which relies heavily on manual processes and limited data analysis, smart factories generate enormous amounts of data from diverse sources—machines, sensors, and software systems. This data provides invaluable insights into every aspect of the production process, from material flow and energy consumption to equipment performance and product quality. Effective data management, analysis, and interpretation are paramount for maximizing the value of this information. This

enables proactive decision-making, leading to improved efficiency and reduced downtime. Think of it as having a comprehensive overview of your entire operation, allowing for near-instantaneous responses to potential problems or changes in demand.
Central to this data-driven approach are technologies like the Internet of Things (IoT). IoT devices, embedded within factory equipment and systems, constantly collect and transmit data, providing a real-time view of the production process. This data, combined with advanced analytics and machine learning algorithms, allows for predictive maintenance—anticipating equipment failures and scheduling maintenance proactively, minimizing costly downtime. Further, this integration enables efficient resource allocation, optimized inventory management, and enhanced traceability throughout the supply chain. The ability to pinpoint problems before they impact production is a substantial advantage over traditional reactive maintenance strategies.
However, implementing smart manufacturing isn't a simple task. It requires a strategic approach, beginning with a clear understanding of your company's needs and goals. A successful transformation demands a phased implementation strategy, starting with pilot projects to test and refine processes before full-scale deployment. This approach allows for incremental improvements and avoids disruptions to existing operations while building expertise within the organization. It's about starting small, learning, and adapting to build a robust and sustainable smart factory environment.
Furthermore, the human element remains crucial. While automation plays a significant role, smart manufacturing isn't about replacing human workers. Instead, it's about empowering them with the tools and information they need to perform their tasks more effectively and efficiently. This means investing in training and development programs, equipping workers with the skills needed to operate and maintain advanced technologies, and creating a work environment that fosters collaboration between humans and machines. A successful transition requires upskilling the workforce, creating a collaborative environment where humans and machines work together seamlessly. Consider this a critical investment in your existing and future workforce.

Ultimately, the success of a smart manufacturing initiative hinges on a holistic approach. It demands seamless integration of various technologies, effective data management, and a clear understanding of the business goals. It's about creating a flexible and responsive production environment that can adapt quickly to changing market demands and optimize resource utilization at every stage. Adopting a well-defined strategy and a phased implementation approach, coupled with investments in training and development, will lay the groundwork for a successful transformation into a truly smart manufacturing operation. This ensures long-term competitiveness and sustainability in today's dynamic and challenging marketplace. The rewards are evident: increased efficiency, improved quality, reduced costs, and enhanced customer satisfaction.

1.2 Benefits of Smart Factories

Enhanced productivity is arguably the most significant advantage. Smart factories leverage automation to drastically reduce downtime, leading to faster production cycles and increased output. Imagine a production line operating continuously, with robots performing repetitive tasks with unwavering precision and speed, far exceeding human capabilities. This translates directly into higher profits and a stronger competitive edge. Think of the implications for meeting fluctuating demands – a smart factory can adjust its output in real time, reacting instantly to market shifts and customer orders. This agility is invaluable in today's dynamic business environment, ensuring you never miss a beat. The efficiency gains ripple through the entire operation, impacting everything from raw material usage to energy consumption.

Beyond the immediate boost in output, smart factories pave the way for substantial cost reductions. Automation minimizes manual labor, reducing labor costs significantly. Furthermore, the precise nature of automated systems minimizes errors, leading to a dramatic decrease in waste and rework. Consider the costs associated with defective products – raw materials, labor, and the time lost in correction. Smart factories significantly reduce these costs, leading to a healthier bottom line. Predictive maintenance, powered by machine learning algorithms analyzing sensor data, minimizes unexpected breakdowns, further reducing operational expenses. Investing in a smart factory might seem expensive initially, but the long-term cost savings are undeniable.

Improved product quality is another major benefit. Automated systems are programmed for precise execution, minimizing the human error that can lead to defects. The data collected by sensors and integrated systems provides real-time feedback, enabling immediate identification and correction of any deviations from the desired quality parameters. Imagine consistently meeting the highest quality standards, leading to increased customer satisfaction and brand loyalty. This improvement extends beyond simple defect reduction; advanced analytics can identify subtle patterns indicating potential quality issues before they escalate, preventing costly recalls and reputational damage. The improved consistency in product quality, from batch to batch, guarantees customer satisfaction and secures a strong position in the market.

Smart factories foster enhanced responsiveness to market demands. By integrating various systems and employing real-time data analysis, they adapt quickly to changing consumer preferences and market trends. This flexibility allows manufacturers to quickly adjust production lines, introduce new products, and respond to urgent customer needs. Consider the advantage of promptly meeting a sudden surge in demand, avoiding lost sales and maintaining a strong competitive posture. This responsiveness is crucial in today's volatile markets, where quick reactions can mean the difference between success and failure. Adaptability is no longer a luxury, but a necessity for long-term survival, and smart factories offer the means to achieve that.

The data-driven nature of smart factories unlocks valuable insights that translate into continuous improvement. The massive amounts of data collected throughout the production process are analyzed to identify bottlenecks, inefficiencies, and areas ripe for optimization. This constant feedback loop allows for continuous improvement, leading to ongoing efficiency gains and cost reductions. Imagine a system learning and adapting over time, constantly striving for optimal performance. This data-driven approach is not just about reactive problem-solving; it's about proactive optimization, anticipating issues before they arise, and pushing the boundaries of productivity and efficiency. The power of data-driven insights is a cornerstone of the smart factory's success, offering a sustainable path to constant improvement.

Beyond the economic benefits, smart factories contribute to a safer and more sustainable manufacturing environment. Automated systems can handle hazardous tasks, reducing workplace risks for human workers. Furthermore, smart factories offer the possibility of optimizing energy consumption and reducing waste, contributing to a more environmentally responsible production process. Consider the ethical implications of providing a safer working environment for employees while minimizing the environmental impact of production. This is more than simply meeting regulations; it's about proactive commitment to responsible manufacturing, enhancing both employee wellbeing and environmental sustainability. It reflects a commitment to a better future.

2 Intelligent Automation Technologies

Robotic systems represent a cornerstone of intelligent automation. These aren't just simple, repetitive machines; advancements allow for sophisticated dexterity and adaptability. Collaborative robots, or cobots, work alongside human employees, enhancing safety and productivity through shared tasks. Consider a paint shop, where a robot applies a perfect coat consistently, while a human worker handles intricate details or quality checks. This synergistic approach maximizes efficiency and reduces errors, benefiting both the production line and the workforce. Implementing these systems requires careful planning; consider the specific needs of your production line, the safety training necessary for human-robot interaction, and the integration with existing infrastructure. Failure to properly plan can result in costly downtime and safety risks. Thorough risk assessment and investment in appropriate safety mechanisms are paramount.

Integrating AI and machine learning (ML) goes beyond simply automating existing processes; it fundamentally transforms how we approach production. ML algorithms, trained on vast datasets of manufacturing data, can predict equipment failures, optimize resource allocation in real-time, and even improve product design. This predictive maintenance, for example, allows for proactive repairs, preventing costly downtime and reducing maintenance expenses. Imagine a scenario where a machine's sensor data indicates a potential bearing failure days before it occurs. This advanced warning allows for scheduled maintenance, avoiding unexpected disruptions and production losses. To successfully implement this technology,

organizations need to invest in data infrastructure capable of collecting and processing large volumes of information. Furthermore, the ability to interpret the ML algorithms' outputs effectively and implement necessary adjustments is critical for maximizing the return on investment.
The Internet of Things (IoT) acts as the nervous system of the smart factory, connecting machines, sensors, and systems into a unified network. This constant flow of data allows for real-time monitoring and control, providing invaluable insights into production processes. Consider a system that monitors the temperature and pressure within a reactor. An anomaly detected by an IoT-connected sensor can trigger an immediate response, preventing potential damage and maximizing productivity. Moreover, this data fuels the AI and ML systems discussed earlier, providing the raw material for predictive analysis and process optimization. Successful implementation requires careful selection of IoT devices, reliable network infrastructure, and robust data security protocols to ensure the integrity and safety of the collected data. Without proper cybersecurity measures, this interconnectedness can become a major vulnerability.
The transition to intelligent automation isn't merely about technology; it necessitates a shift in mindset and processes within your organization. Successfully integrating these technologies requires a comprehensive strategy that addresses not just the technical aspects but also the human element. Training employees to interact with these new systems, fostering a culture of continuous improvement, and adapting workflows to fully leverage the potential of intelligent automation are all crucial aspects. This shift is a journey, not a destination. It demands ongoing investment in training, adaptation, and refinement. A successful implementation requires active participation from all levels of the organization; from management defining strategies and providing resources, to engineers and floor staff contributing their knowledge and insights to optimize processes. Remember, people are an integral part of the success of this transformation.

2.1 Robotics and Automation

Robotics, at their core, are programmable machines designed to perform tasks automatically. In a smart factory setting, they become integral components, handling repetitive, precise, or dangerous operations with far greater speed and consistency than human workers. Consider welding: robots can execute perfect welds continuously, day and night, without fatigue, consistently producing higher quality and minimizing defects. This not only boosts productivity but also enhances overall product quality, a crucial factor in a competitive market. The increased efficiency directly translates into cost savings and improved profitability. Furthermore, the consistent precision of robotic welding eliminates inconsistencies often found in manual processes, leading to more reliable products and reduced waste. The initial investment in robotic systems might seem substantial, but the long-term return on investment (ROI) is often significant, justifying the upfront cost through improved efficiency and reduced operational expenses. The adaptability of robots also deserves consideration, they can be reprogrammed to handle different tasks easily adapting to changing production needs.

Deploying automation technologies, encompassing robotics and other automated systems, goes beyond simply substituting human labor. It's about optimizing entire production lines, creating a seamlessly integrated system where different automated processes cooperate smoothly. Think of a conveyor belt system, seamlessly integrated with robotic arms for picking and placing components, all managed by a central control system. This synchronized orchestration maximizes efficiency and minimizes downtime. Such holistic integration enables faster production cycles, reduced lead times, and increased throughput—all key aspects of a successful modern factory. The effective management of this system requires skilled personnel capable of programming, maintaining, and troubleshooting this complex interconnected machinery. This, therefore, implies a shift in workforce requirements, necessitating training programs focused on maintaining and troubleshooting automated systems. It is a significant change from the traditional manufacturing workforce's skill set.

The selection of the right robotic systems is paramount. It's not a one-size-fits-all solution; the optimal choice depends heavily on the specific tasks, production volume, and budget. Factors to consider include the robot's payload capacity, its reach, its speed, and its precision. Understanding the limitations of each type of robot—industrial robots, collaborative robots (cobots), and SCARA robots—is critical for making an informed decision. Researching different robotic manufacturers and models, carefully comparing their specifications and capabilities, is crucial. Thorough cost-benefit analysis, factoring in installation, maintenance, and training costs, is vital to ensuring the financial viability of the automation project. Don't overlook the crucial role of integration with existing systems, ensuring compatibility and seamless data flow. The overall implementation strategy requires meticulous planning, involving detailed analysis of production processes, careful selection of equipment, and effective project management.

Beyond the immediate gains in efficiency and quality, implementing robotics and automation unlocks further opportunities. Data collected by the robots and other automated systems creates a wealth of information regarding production performance. This data, when analyzed effectively, reveals bottlenecks, inefficiencies, and areas for further optimization. This data-driven approach fosters continuous improvement, leading to ever-increasing productivity and reduced waste. Predictive maintenance, another significant advantage, becomes possible. By analyzing sensor data, potential equipment failures can be anticipated and addressed proactively, reducing downtime and maximizing equipment lifespan. This proactive approach significantly minimizes costly unexpected equipment failures, potentially saving a company a significant amount of money in the long run, increasing productivity, and minimizing disruptions to the production line. Therefore, the integration of robotics and automation is not merely about replacing human workers; it is about creating a dynamic, self-optimizing production environment.

The human element remains crucial. While robots handle repetitive tasks, human workers focus on complex problem-solving, creative design, and strategic decision-making. This synergistic collaboration between humans and robots maximizes the benefits of automation while retaining the irreplaceable value of human ingenuity and adaptability. It is imperative to implement robust safety protocols to ensure worker safety alongside robots, integrating safety features and training programs. Successful implementation demands more than just purchasing robots; it requires a comprehensive strategy encompassing careful planning, skilled implementation, and ongoing monitoring and optimization. Investing in training programs to equip your workforce with the skills to maintain and work alongside these new systems is crucial. This upskilling will empower your employees to fully leverage the potential of your automated systems. The future of manufacturing is not a replacement of humans by machines, but a collaborative partnership between them.

2.2 AI and Machine Learning

Machine learning, a subset of AI, offers powerful tools for enhancing smart factory operations. It goes beyond simple automation, enabling systems to learn from data, adapt to changing conditions, and improve their performance over time. This adaptive capability is crucial in dynamic manufacturing environments where unexpected variations in raw materials, equipment performance, or demand are common. For example, predictive maintenance algorithms, trained on historical equipment data, can anticipate potential failures, scheduling maintenance proactively, minimizing downtime and maximizing uptime. This reduces costly repairs and replacements, improving overall efficiency. Implementing such systems requires careful consideration of data quality, algorithm selection, and integration with existing factory infrastructure.

Consider the potential of machine learning in quality control. By analyzing images and sensor data from the production line in real-time, machine learning models can identify defects with greater speed and accuracy than human inspectors. This results in fewer faulty products reaching the market, improving customer satisfaction and reducing the costs associated with recalls and rework. Imagine the precision achievable with algorithms capable of detecting subtle variations invisible to the naked eye, ensuring consistently high-quality output. Furthermore, the continuous learning aspect of these models means their accuracy improves over time, adapting to new types of defects as they arise. The key here is building a robust data pipeline that captures relevant information from across the factory floor.

AI-powered optimization algorithms can significantly enhance efficiency across the manufacturing process. These algorithms can analyze vast amounts of data – from production schedules and machine performance to energy consumption and material usage – to identify areas for improvement. By simulating different scenarios and evaluating their potential outcomes, they can recommend optimal parameters for maximizing output, minimizing waste, and reducing operational costs. These aren't just theoretical gains; we're talking tangible reductions in energy bills, material waste, and labor costs, translating into a significant increase in profitability. Choosing the right algorithm, however, requires understanding the specific challenges of your production line and the data available to inform the optimization process.

Integrating AI into existing systems can sometimes pose challenges. Legacy systems, lacking the necessary connectivity and data infrastructure, may require significant upgrades to fully leverage the potential of AI-powered solutions. This can involve substantial investment in new hardware, software, and training for personnel. It's vital to plan for these integrations carefully, ensuring compatibility with existing technologies and allowing for a smooth transition. Careful planning minimizes disruption to the ongoing production processes while maximizing the benefits of the upgrade. A phased approach, starting with pilot projects in specific areas before widespread implementation, is often a sensible strategy.

However, successfully implementing AI in a smart factory goes beyond technological considerations. It necessitates a fundamental shift in the way production processes are managed and monitored. Companies must foster a data-driven culture, where data collection and analysis are integral to decision-making at all levels. This includes investing in personnel with the skills to build, deploy, and maintain AI-powered systems, as well as ensuring that the systems integrate seamlessly with existing workforce capabilities. Training employees to effectively interact with and interpret data from these intelligent systems is just as critical as the technological implementation itself. The human element remains crucial for oversight, interpretation, and adaptation in an increasingly automated world. This shift may require a change in company culture to embrace the possibilities and challenges of data-driven decision-making.

The ethical implications of using AI in manufacturing should not be overlooked. Issues such as data privacy, job displacement, and algorithmic bias must be addressed proactively. Transparency in AI systems and ongoing monitoring of their performance are essential to ensuring fairness and accountability. Implementing measures to mitigate potential risks is not just a matter of good practice but also a necessary step to ensure public trust and responsible innovation. It's about building a system that is not only efficient and profitable but also ethically sound. The adoption of AI should enhance, not replace, the importance of human oversight and ethical considerations within the manufacturing process.

2.3 IoT in Smart Factories

The Internet of Things (IoT) forms the nervous system of a smart factory, a vast network of interconnected devices constantly exchanging data. Think of it as a digital circulatory system, pumping information throughout the entire production process. This continuous flow of real-time data empowers unprecedented levels of monitoring, control, and optimization, moving far beyond the capabilities of traditional manufacturing. From sensors embedded in machines to smart cameras tracking product movement, every element contributes to a holistic view of factory operations. This comprehensive data collection isn't just about seeing what's happening; it's about predicting potential problems before they arise.

Predictive maintenance is a game-changer enabled by IoT. Instead of relying on scheduled maintenance routines, which can be costly and disruptive, factories can now anticipate equipment failures. Sensors monitor machine vibrations, temperature fluctuations, and other key performance indicators. When anomalies are detected, the system sends alerts, allowing for timely interventions. This proactive approach minimizes downtime, extends the lifespan of equipment, and significantly reduces maintenance expenses. It's like having a built-in mechanic constantly evaluating your factory's health.

Furthermore, IoT facilitates real-time process optimization. By analyzing the data streams from various sensors, managers can identify bottlenecks, inefficiencies, and areas for improvement. This granular level of data allows for finely-tuned adjustments to production parameters, optimizing resource allocation, and maximizing output. Imagine adjusting conveyor belt speeds based on current production demand, automatically re-routing products based on quality checks, or dynamically adjusting energy consumption to match production needs. This responsiveness isn't possible without the constant flow of information facilitated by the IoT.

Beyond individual machines, IoT connects the entire factory ecosystem. This interconnectedness is vital for effective supply chain management. Tracking materials and components from suppliers to the factory floor, monitoring inventory levels in real time, and coordinating logistics creates efficiencies throughout the supply chain. This integrated approach reduces delays, minimizes waste, and ensures a smoother, more reliable supply of materials. It's about seeing the entire process as a seamless whole, not just isolated parts.

Data security remains paramount in the context of IoT. The vast amount of data collected by connected devices makes smart factories potential targets for cyberattacks. Therefore, robust security measures are crucial. Implementing secure protocols and encryption techniques is non-negotiable. Regular security audits, firewall protection, and intrusion detection systems are essential to safeguarding sensitive data and protecting the factory's operational integrity. Remember, the interconnected nature that gives IoT its power also creates vulnerabilities that need to be addressed proactively.

Implementing IoT in a smart factory requires careful planning and execution. Start with a clear definition of your goals and objectives. What are you hoping to achieve with IoT integration? Are you focusing on predictive maintenance, process optimization, or supply chain improvements? Once your goals are clearly defined, you can begin selecting the appropriate sensors, software, and communication protocols. Consider starting with a pilot program in a limited area of the factory before deploying IoT across the entire operation. This phased approach allows you to test the system, identify potential problems, and refine your implementation strategy before scaling up.

Finally, don't overlook the importance of workforce training and engagement. A successful IoT implementation relies on employees understanding the system and being able to use the data it provides. Training programs should cover data analysis, interpretation, and problem-solving techniques. Ensure the system is user-friendly and intuitive to encourage active participation and adoption. Remember, the human element remains a critical component of any successful smart factory initiative. The benefits of IoT are realized only when the data is effectively used to inform and improve human decision-making. By embracing this collaborative approach, factories can harness the full potential of IoT to achieve greater efficiency, productivity, and profitability.

3 Streamlining Production Processes

Adopting lean manufacturing principles is fundamental to streamlining your production processes. This involves identifying and eliminating waste in all its forms—from excess inventory and unnecessary movements to defects and overproduction. Imagine a factory floor where every action contributes directly to value creation; that's the lean ideal. Visualize the flow of materials, the smooth transitions between workstations, the minimal wait times. Implementing kanban systems, 5S methodologies, and value stream mapping can drastically improve efficiency. These tools help visualize your workflow, identify bottlenecks, and optimize material flow. By meticulously analyzing each step of your production process, you'll uncover opportunities for significant improvements. Remember, even small incremental changes can accumulate to substantial gains over time. Continuous improvement, or Kaizen, is at the heart of lean thinking. Embrace this philosophy, and you'll foster a culture of constant refinement and optimization within your factory.

Process optimization techniques go beyond simply removing waste. It's about systematically analyzing your production processes, identifying areas for improvement, and implementing targeted solutions. This involves utilizing tools such as Six Sigma, which emphasizes minimizing defects and improving process capability. Consider using process mapping to create a detailed visual representation of your workflow, then carefully scrutinize each step for potential bottlenecks, redundancies, or opportunities for automation. Data analysis, discussed in a later chapter, is critical here;

it helps to pinpoint areas where improvements will yield the greatest impact. Implementing robust process control measures ensures consistency and predictability, reducing variability and potential defects. Remember, effective process optimization is an iterative process; constant monitoring, evaluation, and adaptation are essential for sustained improvements. The goal is not just to achieve a single improvement, but to build a culture of continuous process refinement.

Digital twins offer incredible potential for optimizing your processes. A digital twin is a virtual replica of your physical production environment, allowing you to simulate different scenarios, test new strategies, and predict potential problems before they occur. Imagine being able to test a new production line layout virtually, identifying potential bottlenecks and inefficiencies without disrupting your actual production. This technology enables predictive maintenance, allowing you to anticipate equipment failures and schedule maintenance proactively, minimizing downtime and maximizing productivity. By simulating different scenarios, you can optimize production parameters, identify potential risks, and explore different configurations. Digital twins empower informed decision-making, reducing risk and accelerating the implementation of effective improvements. Furthermore, it promotes collaboration by allowing different teams to visualize and interact with the same model, fostering a shared understanding of the production process.

Integrating your supply chain is crucial for efficient production. Streamlining the flow of materials and information from your suppliers to your factory floor reduces lead times and minimizes disruptions. Effective supply chain management involves establishing strong relationships with your suppliers, implementing robust inventory management systems, and utilizing technologies such as RFID tracking for enhanced visibility. Real-time tracking allows for proactive management of potential disruptions, enabling you to adjust your production schedule accordingly. Close collaboration with your suppliers fosters transparency and ensures that materials arrive on time and in the right quantities. Think about using collaborative platforms to share real-time data, enabling everyone in the supply chain to anticipate and respond efficiently to changing demands.

Data analytics provides critical insights into the effectiveness of your production processes. By collecting and analyzing data from various sources—including your machinery, your supply chain, and your quality control systems—you can gain a comprehensive understanding of your operations. Using business intelligence tools, you can identify trends, patterns, and areas for improvement. Consider key performance indicators (KPIs) like production efficiency, defect rates, and downtime. Analyzing these metrics allows you to pinpoint bottlenecks, assess the effectiveness of implemented changes, and make data-driven decisions to continuously improve your operations. The insights gleaned from this data-driven approach enable proactive adjustments, leading to increased efficiency, reduced costs, and higher quality. The ability to analyze vast datasets provides a level of understanding that would be impossible through manual observation alone. This is fundamental to creating a truly smart factory.

3.1 Lean Manufacturing Principles

Eliminating waste is the cornerstone of lean manufacturing. It's not just about reducing materials; it's about identifying and removing all forms of waste throughout the entire production process. This includes things you might not immediately consider, like excessive motion by workers, unnecessary inventory, or even waiting time. By meticulously analyzing each step, you can pinpoint areas ripe for improvement, optimizing efficiency and reducing costs dramatically.

Think of it like this: Imagine a perfectly choreographed dance. Each movement is precise, each step flows seamlessly into the next. That's the ideal lean manufacturing process – a smooth, efficient flow of materials and actions, eliminating any unnecessary steps. This requires a deep understanding of the value stream, which is the sequence of all actions required to bring a product from concept to customer.

The 5S methodology provides a structured approach to this process. Seiri (sort), Seiton (set in order), Seisō (shine), Seiketsu (standardize), and Shitsuke (sustain) form a powerful framework for workplace organization. Implementing 5S isn't just about tidiness; it's about creating a more efficient, safer, and ultimately more productive workspace. By eliminating clutter and organizing tools and materials efficiently, you'll reduce wasted time searching for items and prevent accidents. The sustained improvement built into 5S fosters a culture of continuous improvement.

Visual management tools, such as Kanban systems, play a crucial role in lean manufacturing. These systems provide a real-time visual representation of the workflow, allowing for quick identification of bottlenecks and potential problems. For instance, Kanban boards display the status of each task or product, enabling teams to immediately spot delays and address them proactively, maintaining a consistent and predictable flow. This visibility fosters quicker decision-making and supports faster responses to changing demands, leading to a more agile production system.

Value stream mapping is another critical tool. This technique involves visually mapping out the entire process, from raw materials to finished product, allowing teams to identify non-value-added activities – those steps that don't add value to the final product. By identifying and eliminating these steps, you significantly streamline your process and improve overall efficiency. This methodical approach ensures that every action contributes directly to the creation of a valuable product, minimizing wasted time and resources. The process reveals hidden inefficiencies, offering a pathway to streamlined workflow.

Just-in-time (JIT) inventory management is a core principle of lean manufacturing. This strategy focuses on receiving materials only when they are needed, minimizing storage space and the risk of obsolescence. This requires strong coordination with suppliers and precise production scheduling, optimizing your supply chain. The reduction in storage costs and the minimization of waste are significant advantages, improving profitability and contributing to the agile nature of the system. JIT reduces warehousing needs and allows you to focus on what matters most – providing a high-quality product efficiently.

Continuous improvement, often referred to as Kaizen, is the philosophy driving lean manufacturing's long-term success. It's about embracing small, incremental changes over time, constantly seeking ways to enhance efficiency and eliminate waste. This isn't a one-time event but an ongoing process of refinement, involving every member of the team, ensuring that improvements are integrated into the fabric of the business. This iterative process requires collaboration, and it's through this constant improvement that lean manufacturing truly shines, optimizing the process over time. Implementing lean manufacturing is a journey, not a destination. It requires dedication, a commitment to continuous improvement, and a willingness to challenge existing processes. However, the rewards are substantial. By consistently applying these principles, businesses can significantly improve their efficiency, reduce costs, and enhance overall productivity, gaining a considerable competitive edge in today's fast-paced market. The benefits go beyond simply improving the bottom line; lean manufacturing fosters a culture of teamwork and continuous improvement, leading to a more engaged and satisfied workforce.

3.2 Process Optimization Techniques

Identifying bottlenecks is paramount. Analyze your production line meticulously, examining each step for inefficiencies. Look for areas with excessive waiting times, redundant actions, or high defect rates. Employ tools like value stream mapping to visually represent your process and pinpoint areas ripe for improvement. Don't overlook the seemingly small details; often, cumulative minor inefficiencies create significant overall losses. A thorough analysis, possibly involving time-motion studies, provides concrete data for targeted intervention.

Implementing lean principles directly addresses many common production issues. By systematically eliminating waste – be it in the form of overproduction, waiting, transportation, unnecessary processing, excess inventory, unnecessary motion, or defects – you drastically improve efficiency and throughput. Kaizen events, short focused improvement projects involving the entire team, are effective for identifying and resolving these issues incrementally, fostering a culture of continuous improvement within the factory. The key is to embed lean thinking into the factory's DNA.

Process mapping offers a clear visual representation of your workflow. Detailed process maps, often using flowcharts or swim lane diagrams, allow you to see the sequence of tasks, identify hand-offs, and understand the flow of materials and information. This is crucial for detecting redundancies, delays, or points of friction that might be causing inefficiencies. Regularly reviewing and updating these maps ensures that your process optimization efforts remain relevant and effective as your operation evolves. Software tools are increasingly available to automate the creation and maintenance of these maps.

Data analytics plays a crucial role. Modern smart factories generate massive amounts of data from various sources, including sensors, machines, and ERP systems. By using this data effectively, you can gain valuable insights into process performance, identify patterns, and predict potential problems. Real-time monitoring dashboards, coupled with advanced analytics techniques like predictive maintenance algorithms, can greatly enhance your ability to prevent downtime and maintain optimal performance. Investing in data analysis skills within your team is vital.

Simulation and digital twin technology allows for the testing and optimization of changes without disrupting the actual production line. Creating a virtual representation of your factory enables you to model various process improvements and predict their impact on efficiency, throughput, and overall performance. This reduces the risk of implementing costly changes that may prove ineffective or even detrimental. The use of simulation software offers a safe and controlled environment for experimentation and optimization.

Automation technologies, used judiciously, enhance process optimization. While complete automation isn't always feasible or economically viable, strategically automating specific tasks or processes can significantly improve efficiency and reduce human error. This should be driven by data analysis, identifying the tasks where automation provides the greatest return on investment. Choosing the right automation technology, and carefully integrating it into existing workflows, is crucial for success. Remember, this isn't about replacing workers, but augmenting their capabilities.

Consider the human element. Employee involvement is essential for successful process optimization. Empower your workforce by providing them with training, the tools and data they need to identify and address process issues, and create a collaborative work environment where suggestions are valued and acted upon. A team that feels invested in the optimization process will be more proactive in identifying and solving problems, leading to more effective and lasting improvements. It's all about building trust and teamwork.

Continuous improvement should be an ongoing commitment, not a one-time project. Regularly review key performance indicators (KPIs), track your progress, and continually seek out opportunities for further optimization. This requires a cultural shift towards continuous improvement, where everyone in the organization is focused on identifying and addressing inefficiencies. Regularly scheduled reviews and adjustments allow you to adapt to changing circumstances, technological advancements, and market demands, maintaining a competitive edge. This continuous cycle drives sustainability and success.

3.3 Digital Twins and Simulation

Digital twins offer a powerful tool for simulating and optimizing your factory's operations before implementing changes in the real world. Imagine having a virtual replica of your entire production line, complete with every machine, robot, and conveyor belt. This is the essence of a digital twin. You can use this virtual environment to test new processes, predict potential bottlenecks, and even explore the impact of different equipment layouts without disrupting your actual production. This allows for proactive problem-solving and avoids costly mistakes during implementation. It provides a safe space to experiment, a sandbox where failure is simply a learning opportunity, leading to more efficient and effective outcomes.

The creation of a digital twin requires meticulous data collection. Every aspect of your factory, from individual machine performance data to the overall workflow, needs to be accurately represented in the virtual model. This often involves integrating data from various sources, including sensors, PLCs, and ERP systems. Data accuracy is paramount; garbage in, garbage out. The twin's effectiveness hinges on the reliability and completeness of this input data, ensuring that the digital world mirrors the physical one as faithfully as possible. Consequently, investing in robust data acquisition systems and cleaning procedures pays enormous dividends in achieving an effective and useful digital twin. Think of it as building a highly detailed blueprint, ensuring every component is meticulously represented.

Simulation, powered by the digital twin, allows you to run countless scenarios. What happens if you increase production speed by 15%? What if a machine fails unexpectedly? The simulation can provide predictive analytics and highlight potential issues before they even arise. By experimenting with different parameters in the digital twin, you can identify optimal settings, significantly reducing waste and optimizing overall efficiency. You can also test different scenarios for maintenance schedules, identifying the most cost-effective strategies and mitigating risks of downtime. This predictive capability allows for proactive planning and a smoother, more productive operation. It shifts your focus from reactive problem-solving to proactive optimization.

One crucial aspect is the selection of the appropriate simulation software. Many options exist, ranging from general-purpose simulation tools to specialized software tailored for manufacturing environments. The choice depends on factors such as the complexity of your production processes, your budget, and the specific functionalities you require. Consider factors like ease of use, data integration capabilities, and the software's ability to handle large datasets. It's worth taking the time to thoroughly research and compare different options before making a decision. A robust and user-friendly software platform will streamline your simulation workflow and ensure that the process is efficient and effective. Don't underestimate the importance of proper software selection; it can make or break your digital twin initiative.

Beyond just process optimization, the digital twin can help you optimize resource allocation. You can model the impact of changes in staffing levels, material flow, and energy consumption. This can lead to significant cost savings and improved sustainability. For example, simulating different energy-saving measures can identify the most impactful strategies with minimal disruption. The ability to model a variety of 'what if' scenarios is invaluable, potentially transforming your factory's operational efficiency and contributing to a more environmentally responsible production process. It allows a deeper understanding of resource utilization and can uncover hidden areas for efficiency enhancement.

Remember, the digital twin is a living model. As your factory evolves, so should your digital twin. Regular updates are crucial to ensure the virtual model accurately reflects the real-world state. This includes incorporating data from new equipment, process changes, and any other modifications to your production line. The iterative nature of maintaining a digital twin is key to ensuring its continued relevance and usefulness. It's not a static model; it's a dynamic tool that grows and adapts alongside your factory's evolution. Regular updates ensure that the simulations remain relevant and the insights generated remain accurate and impactful.

Finally, don't view the digital twin as an isolated project. It should be integrated into your broader digital transformation strategy. The data generated by the digital twin can inform other aspects of your factory's operations, such as predictive maintenance, quality control, and supply chain management. This interconnectedness is crucial for extracting maximum value from your digital twin investment. It is part of a larger picture, synergizing with other smart factory technologies to create a holistic, optimized, and adaptive production environment. The digital twin's value extends far beyond simply simulating processes; it forms a core component of a comprehensive smart factory strategy.

3.4 Supply Chain Integration

Effective supply chain integration is paramount for a truly smart factory. It's not merely about connecting disparate systems; it's about creating a seamless, responsive network that optimizes the flow of information and materials from raw material sourcing to final product delivery. This holistic approach minimizes delays, reduces waste, and enhances overall efficiency, ultimately boosting profitability and competitiveness. Ignoring this crucial aspect undermines the very potential of a smart factory.

Consider the implications of a poorly integrated supply chain. Delays in receiving crucial components can halt production lines, resulting in lost revenue and dissatisfied customers. Lack of real-time visibility into inventory levels leads to overstocking or shortages, tying up capital and potentially disrupting production schedules. Inefficient communication between different stages of the chain creates bottlenecks and inefficiencies, wasting valuable time and resources. These are not mere theoretical concerns; they are common pitfalls that can easily derail even the most ambitious smart factory initiatives.

To achieve successful supply chain integration, you need a robust, interconnected system. This involves implementing advanced technologies such as RFID tracking, blockchain for enhanced transparency and traceability, and cloud-based platforms for real-time data sharing and collaboration. Furthermore, it requires a strategic approach that considers all aspects of the supply chain, from supplier relationships to logistics and warehousing. Investing in these technologies is essential for realizing the full potential of a smart factory. It's a necessary investment, not an optional extra.

Building strong relationships with suppliers is crucial. Collaboration fosters trust and open communication, enabling proactive problem-solving and efficient information sharing. This includes sharing real-time production data with suppliers to optimize order quantities and delivery schedules. Joint planning and forecasting can help anticipate potential disruptions and mitigate their impact. Trust and communication aren't just buzzwords; they're the cornerstone of a successful integrated supply chain, promoting efficiency, flexibility, and innovation throughout the entire process.

Logistics and warehousing play a vital role in efficient supply chain management. Implementing automated warehousing systems can significantly reduce handling times, improve accuracy, and minimize storage costs. Advanced route optimization software can streamline delivery schedules, reducing transportation costs and delivery times. Implementing real-time tracking and monitoring ensures accurate inventory control and proactive identification of potential delays or disruptions. These tools are not luxuries but necessities to ensure that goods are in the right place at the right time, reducing waste and enhancing responsiveness.

Data analytics are indispensable in optimizing supply chain operations. By leveraging advanced data analytics techniques, you can gain valuable insights into various aspects of your supply chain, such as identifying bottlenecks, predicting demand fluctuations, and optimizing inventory levels. This allows for proactive decision-making and continuous improvement, leading to improved efficiency and cost reduction. Data-driven insights are your edge; using them effectively isn't just advantageous but crucial for gaining a competitive edge.

Successful integration requires a well-defined strategy, encompassing clear objectives, a phased implementation plan, and robust change management. Don't attempt a complete overhaul overnight; begin with a pilot project to test and refine processes before scaling up. Invest in training and development for your employees to ensure seamless transition and adoption of new systems and processes. Careful planning and execution ensure a smooth transition to a smarter, more integrated supply chain.

Finally, remember that supply chain integration is an ongoing process, not a one-time event. Continuous monitoring, evaluation, and improvement are critical for ensuring optimal performance and adaptation to changing market conditions. Regular reviews and adjustments to your strategy based on performance data will ensure your supply chain remains aligned with your overall business objectives. This continuous improvement is the key to remaining competitive and efficient in the ever-evolving world of manufacturing. It's not just about building a smart factory, it's about keeping it smart.

3.5 Data Analytics for Improvement

Analyzing your factory's data isn't just about collecting numbers; it's about uncovering hidden opportunities for improvement. Think of it as a treasure hunt, where the treasure is increased efficiency and reduced costs. The right tools and techniques can transform raw data into actionable insights, driving significant positive change within your operations. Identifying bottlenecks, predicting equipment failures, and optimizing resource allocation—these are just a few of the ways data analytics can enhance your bottom line. The key is to use this data to make intelligent decisions, moving away from reactive problem-solving to a more proactive and strategic approach to management.

Data visualization plays a critical role in this process. Charts and graphs make complex data sets easily understandable, revealing patterns and trends that might otherwise go unnoticed. A well-designed dashboard can provide a real-time overview of key performance indicators (KPIs), allowing managers to quickly identify areas needing attention. For instance, a sudden spike in machine downtime might indicate a looming maintenance issue, allowing for preventative action before a full-blown breakdown occurs. Such early warning systems are invaluable in a fast-paced manufacturing environment. The visualization should be tailored to your specific needs; there's no one-size-fits-all solution. Experimentation is key to finding the optimal approach for your business.

Predictive maintenance is another powerful application of data analytics. By analyzing historical data on machine performance, you can predict when equipment is likely to fail. This allows for scheduled maintenance during less disruptive times, minimizing downtime and maximizing productivity. This proactive approach prevents costly emergency repairs and extends the lifespan of your equipment. Consider factors like vibration levels, temperature fluctuations, and energy consumption; each data point provides a piece of the puzzle. The more data you collect, the more accurate your predictions will be. This doesn't mean replacing human expertise; rather, it enhances it by providing informed decision-making support.

Beyond predictive maintenance, data analytics can optimize your entire production process. Analyzing production data, for instance, can reveal bottlenecks and inefficiencies that are otherwise hidden. Are certain production lines consistently slower than others? Are there specific tasks that are taking up an excessive amount of time? Identifying these bottlenecks allows you to focus your improvement efforts on the areas with the greatest potential for impact. This targeted approach ensures that your resources are used effectively, maximizing the return on investment in your improvement initiatives. You will significantly reduce waste, increase productivity, and streamline your production process, improving overall efficiency.

Furthermore, data analytics can improve quality control. By tracking defects and identifying their root causes, you can implement preventative measures to reduce future errors. This proactive approach reduces waste, improves product quality, and enhances customer satisfaction. This isn't solely about detecting defects after they've occurred; it's about understanding the processes and equipment that contribute to those defects in the first place. By addressing these underlying issues, you create a system that is inherently more robust and less prone to errors. Real-time data monitoring enables swift responses to emerging quality concerns.

Finally, remember that data analytics is an iterative process. It's not a one-time fix but rather a continuous cycle of data collection, analysis, and improvement. Start with a clearly defined goal, collect the relevant data, analyze the results, and implement improvements. Then, repeat the process, constantly refining your approach based on the insights you gain. Embrace experimentation and don't be afraid to adjust your strategy as needed. The key is to be flexible and adapt to the evolving needs of your business. The value of data analytics lies in its ability to provide a constant stream of valuable improvements, leading to continuous growth and enhanced competitiveness within the dynamic manufacturing landscape. The power lies in the iterative nature of the process, leading to continuous enhancement and optimization.

4 Cybersecurity in Smart Factories

The interconnected nature of smart factories, while offering unprecedented efficiency, introduces significant cybersecurity vulnerabilities. Think of it like this: your factory is now a complex web, each machine a node, constantly communicating and sharing data. A single weak point, a single compromised device, can unravel the entire system, leading to production downtime, data breaches, and even physical damage. This isn't theoretical; attacks are happening now, targeting everything from industrial control systems to the very software controlling robotic arms. The consequences can be devastating, ranging from financial losses to reputational damage, and even compromising sensitive customer information. Ignoring cybersecurity in this context isn't an option; it's a recipe for disaster.

Protecting your smart factory requires a multi-layered approach. It's not about one magic bullet solution, but rather a robust strategy that addresses every potential entry point. Start with a comprehensive risk assessment. Identify your most critical assets – those systems whose failure would have the most significant impact – and prioritize their protection. Consider physical security too; unauthorized access to your facility can be just as damaging as a cyberattack. Implementing robust access control systems, regular security audits, and thorough employee training are vital initial steps. Remember, your employees are often the weakest link, so make sure they understand the importance of secure passwords, phishing awareness, and reporting suspicious activity. This holistic approach helps prevent threats from multiple angles.

Network segmentation is crucial. Don't let your entire factory operate on a single network; segment it into smaller, isolated networks to limit the damage from a successful breach. This means separating your IT network from your operational technology (OT) network, and further subdividing these networks based on function and sensitivity of the data. Each segment requires its own security measures, tailored to its specific needs. This isn't just about firewalls; it's about carefully controlling the flow of information within your factory, ensuring that only authorized devices and users can access specific parts of your system. This segmented approach reduces the risk of a single attack cascading through the entire operation.

Investing in advanced security technologies is non-negotiable. This means implementing intrusion detection and prevention systems (IDPS) to monitor network traffic for malicious activity. Consider employing endpoint detection and response (EDR) solutions to protect individual devices from malware and other threats. Regularly updating your software and firmware is essential, patching vulnerabilities before attackers can exploit them. A robust vulnerability management program should be a cornerstone of your cybersecurity strategy. Think of this like regularly servicing your machinery; neglecting it will lead to breakdowns, just as neglecting software updates can leave your systems vulnerable.

Finally, remember that cybersecurity isn't a one-time event; it's an ongoing process. Regular security assessments, employee training, and system updates are critical to maintain a strong defense. Develop an incident response plan – a detailed strategy for handling a security breach, minimizing its impact, and recovering quickly. This plan shouldn't just sit on a shelf; it should be regularly tested and updated to reflect changes in your environment and the evolving threat landscape. A well-rehearsed incident response plan is the difference between a contained incident and a catastrophic failure. The ongoing commitment to security is not merely a cost; it's an investment protecting the long-term health and profitability of your smart factory. The cost of inaction far outweighs the cost of proactive protection.

4.1 Vulnerabilities and Threats

The interconnected nature of smart factories, while offering immense productivity gains, introduces significant cybersecurity vulnerabilities. A single compromised system can cascade failures throughout the entire operation, causing production halts, data breaches, and even physical damage to equipment. This presents a serious threat to not only the profitability but also the safety of the entire facility. Consider, for example, a malicious actor gaining control of a robotic arm—the consequences could range from minor disruptions to catastrophic accidents.

Unlike traditional factories, smart factories rely heavily on networked systems, creating a vast attack surface. The Internet of Things (IoT) devices, while providing valuable data, often lack robust security features, making them easy targets for intrusion. These vulnerabilities, coupled with the increasing sophistication of cyberattacks, pose a significant challenge to smart factory security. The reliance on cloud-based data storage and analysis also introduces external vulnerabilities, increasing the risk of data breaches and intellectual property theft.

Imagine a scenario where a ransomware attack cripples a smart factory's control systems. Production grinds to a halt, incurring substantial financial losses. Beyond direct financial implications, such attacks can also damage a company's reputation, leading to loss of customer trust and long-term market share decline. The impact extends beyond monetary loss, affecting supply chains and potentially endangering employees depending on the nature of the compromised system. These disruptions can impact deadlines and lead to severe penalties from clients.

Effective security requires a multi-layered approach. This is not a one-size-fits-all solution; a tailored approach must be taken considering the specific vulnerabilities and threats unique to each factory's setup and operational requirements. Regular software updates and security patches are crucial, yet frequently overlooked. Out-of-date systems, particularly those responsible for critical processes, represent significant points of weakness. Neglecting software updates can easily expose a factory to known vulnerabilities.

Intrusion detection systems (IDS) and intrusion prevention systems (IPS) should be deployed to monitor network traffic for suspicious activity and proactively block potential threats. These systems, when properly configured, are capable of detecting and responding to various threats such as malware, denial-of-service attacks, and unauthorized access attempts. Investing in advanced security measures isn't merely an expense but a crucial investment for long-term operational stability and profitability. Failure to properly secure systems can have very serious consequences.

Employee training is another critical aspect often underestimated. A well-trained workforce is better equipped to identify and report suspicious activities. Regular security awareness training should educate employees about phishing attempts, social engineering tactics, and the importance of strong password hygiene. Humans represent a significant weak point within a cybersecurity strategy. Simple human errors, like clicking on malicious links, can compromise the entire network. It is, therefore, crucial to mitigate this risk through employee education.

Beyond technical measures, robust security policies and procedures are essential. These policies should define roles and responsibilities, access control measures, incident response plans, and regular security audits. A clearly defined incident response plan is essential for containing the damage in case a breach does occur. This plan should outline procedures for isolating infected systems, containing the spread of malware, restoring systems from backups, and collaborating with law enforcement if necessary. Such plans are not merely "theoretical" documents but vital tools that should be regularly tested and updated.

The integration of cybersecurity into the initial design and implementation phases of smart factory projects is paramount. Building security into the system from the ground up is far more effective and cost-efficient than retrofitting security measures later on. A proactive approach emphasizes prevention over remediation. By prioritizing security from the outset, organizations can significantly reduce their risk exposure. Failing to do so can lead to costly and disruptive remediation efforts further down the line. This proactive approach should extend to both hardware and software choices, ensuring each component aligns with the factory's overall security strategy.

Finally, staying informed about the latest cybersecurity threats and vulnerabilities is essential for maintaining a strong security posture. Regularly updating security protocols and incorporating emerging technologies such as blockchain and advanced authentication methods, allows a smart factory to stay ahead of increasingly sophisticated attacks. A well-informed and adaptable security approach is a continuous process, not a single event or project. The ever-evolving nature of cyber threats necessitates consistent vigilance and adaptation to ensure long-term security. This vigilance and adaptability are key to mitigating the risks associated with vulnerabilities and threats in the smart factory environment.

4.2 Security Measures and Protocols

Implementing robust cybersecurity is paramount in smart factories. The interconnected nature of these facilities, with their numerous sensors, robots, and control systems constantly exchanging data, creates a vast attack surface. A single point of vulnerability can compromise the entire operation, leading to production downtime, data breaches, and significant financial losses. Therefore, a multi-layered approach to security is essential.

This requires a proactive strategy rather than simply reacting to incidents. Regular security assessments are vital. These should identify weaknesses in your systems before malicious actors can exploit them. Think of it like a medical checkup; preventative measures are far more cost-effective than emergency treatment. Employ penetration testing to simulate real-world attacks, allowing you to identify vulnerabilities and strengthen your defenses. Don't just focus on the obvious entry points; look for less apparent ways a hacker might penetrate your system.

Network security is the foundation of your smart factory's digital fortress. A strong firewall is the first line of defense, acting as a gatekeeper, controlling access to your network. But a firewall alone isn't enough. Implement intrusion detection and prevention systems (IDPS) to monitor network traffic for suspicious activity. These systems can identify and block malicious attempts to access your network before they can cause harm. Regular updates to your firewall and IDPS are crucial; think of it as regularly updating the software on your computer—it keeps your defenses strong and current.

Data security is another critical element. Your factory generates a massive amount of sensitive data, from production parameters to proprietary designs. Robust encryption protocols are crucial to safeguarding this information, both in transit and at rest. Implement access control measures to ensure only authorized personnel can access sensitive data, using a principle of least privilege – each user only having the access they absolutely need to perform their job. Regular audits of access logs will help track who has accessed what, helping to detect any unauthorized activity.

Consider adopting a zero-trust security model. This framework assumes no implicit trust, requiring verification at every access point. Multi-factor authentication adds an extra layer of security, requiring users to provide multiple forms of identification to gain access. This makes it significantly more difficult for unauthorized users to gain access, even if they manage to obtain a password.

Beyond network and data security, consider the physical security of your equipment. Robust access controls to your factory floor, along with surveillance systems, can deter unauthorized access and sabotage. Regular maintenance checks on physical security measures are also essential, ensuring that your physical barriers remain strong and reliable.

Employee training plays a crucial role. Regular security awareness training should educate employees on common threats like phishing emails and social engineering attacks. Equipping your staff with the knowledge to recognize and respond to these threats reduces the risk of human error, a common vector for attacks. Regular simulated phishing campaigns can test employees' vigilance, highlighting areas where additional training is needed. Furthermore, establish clear incident response procedures, providing step-by-step guidance on how to handle security incidents, ensuring a swift and effective response.

Finally, remember that cybersecurity is an ongoing process, not a one-time task. Regular updates to software and hardware, continuous monitoring, and ongoing employee training are all essential to maintaining a strong security posture. A comprehensive security plan is crucial, covering all aspects of your smart factory's operations and actively adapting to the constantly evolving threat landscape. Staying informed about the latest threats and vulnerabilities will allow you to proactively strengthen your defenses and protect your valuable assets. The cost of inaction far outweighs the investment in robust security measures. Ignoring these protocols risks everything you've built. Consider this your final precaution – your last defense. Your factory's future depends on it.

5 Human-Robot Collaboration

Human-robot collaboration isn't just about robots doing human jobs; it's about creating a synergistic partnership. Think of it as an orchestra, where robots handle the repetitive, precise tasks, freeing up human workers for more creative, problem-solving roles. This new division of labor maximizes efficiency and improves overall output. This requires a careful consideration of workspace design and safety procedures.

Designing a safe and effective collaborative workspace isn't simply about placing robots amongst people. It needs a holistic approach, starting with a thorough risk assessment. This assessment must consider the specific robots, the tasks they'll perform, and the workers involved. Following regulations and industry best practices is critical; companies shouldn't take shortcuts. Implementing safety features like emergency stops and sensors is paramount, and ongoing monitoring is crucial to adapt to evolving needs. Remember, worker safety should be the top priority in any collaborative environment.

The successful integration of robots into a factory necessitates a well-trained workforce. Employees need to understand how to operate and maintain the robots, troubleshoot any problems that may arise, and work alongside them safely and efficiently. This demands a comprehensive training program that covers both theoretical knowledge and hands-on practice. Regular refresher courses and advanced training opportunities are equally important to keep workers up-to-date with technological advancements and best practices. Investing in your workforce is an investment in the success of your smart factory.

Collaborative robots, or cobots, are designed for direct interaction with humans. Unlike traditional industrial robots, which are typically isolated in cages, cobots are designed to work alongside humans safely, sharing the same workspace. Cobots are typically smaller, lighter, and more flexible than their industrial counterparts. Their key feature is their inherent safety mechanisms, preventing injuries during unplanned collisions. Cobots are particularly well-suited for tasks that require both human dexterity and robotic precision, for instance, assembly tasks or quality inspection. Their integration can streamline processes, enhancing productivity.

However, implementing cobots requires careful planning. You need to identify tasks suitable for cobots, integrating them into existing workflows seamlessly. This isn't a simple plug-and-play situation. You need to train your staff, adjust production lines, and monitor for unexpected issues. This might involve initial production slowdowns and increased costs; don't underestimate the time or resources this requires. Thorough planning, effective training, and a robust support system are pivotal to the successful use of cobots.

Successful human-robot collaboration hinges on effective communication and coordination. Humans and robots need to "talk" to each other, sharing information about their actions and status. This often involves integrating sensors and software systems that enable real-time data exchange. This could include sensors to detect the presence of humans, and sophisticated software to coordinate movements to prevent collisions. Effective communication ensures that the human-robot team works efficiently and safely, optimizing overall productivity. This sophisticated communication should be planned and implemented from the initial stages of robot integration.

The shift towards human-robot collaboration represents a significant change in the manufacturing landscape. It's not just about replacing human workers with robots; it's about redefining roles, creating new opportunities, and fostering a more efficient and safer work environment. This requires a proactive approach to change management, where workers feel supported and involved in the transition. Effective communication and transparency are essential to address anxieties and foster a positive attitude toward technological advancements. Training programs should emphasize the benefits of human-robot collaboration, highlighting the new skills workers will acquire and the potential for career advancement. It's about empowering the workforce to embrace this new era of manufacturing.

Beyond immediate practical implications, the success of human-robot collaboration also hinges on careful consideration of ethical implications. As robots increasingly take on more complex tasks, it's important to ensure that their use is aligned with ethical principles and values. This includes ensuring fair treatment of workers, considering the potential impact on employment, and promoting responsible innovation. This requires a collaborative effort, involving stakeholders from across the manufacturing sector, to define guidelines and best practices that promote a just and equitable transition toward a future where humans and robots work side-by-side. Considering the ethical dimensions from the outset will facilitate a smoother and more responsible implementation of human-robot collaborations.

5.1 Ergonomics and Safety

Designing safe and ergonomic workspaces in smart factories is crucial. It's not just about following regulations; it's about fostering a productive and healthy environment where human-robot collaboration thrives. This involves careful consideration of workstation design, reducing repetitive movements, and minimizing physical strain. Think about implementing adjustable workstations that allow employees to comfortably switch between sitting and standing, reducing back pain and fatigue often associated with prolonged static postures. Investing in this kind of equipment isn't just about employee well-being; it's a direct investment in increased productivity and reduced absenteeism.

Proper lighting is often overlooked, but it significantly impacts worker well-being and safety. Insufficient or poorly placed lighting can lead to eye strain, headaches, and even accidents. Ensure your factory uses a lighting system designed to minimize glare and shadows, optimizing visibility for all tasks. Consider using natural light where possible, as it's been shown to improve mood and alertness. This is especially important in areas with high levels of automation, where monotonous tasks can lead to fatigue. For areas with automated machinery, good lighting ensures employees can properly monitor processes and react swiftly if needed, preventing accidents.

The integration of robots and humans necessitates specific safety protocols. These shouldn't be an afterthought; they must be integral to the design and implementation phase. Robust safety mechanisms like emergency stop buttons, light curtains, and pressure-sensitive sensors are essential to prevent accidents. Regular maintenance and inspection of these safety features are just as crucial. Remember, these are not just compliance measures; they are investments in preventing potentially devastating injuries and production downtime. Furthermore, well-designed safety systems can increase worker confidence in the automation systems, allowing for smoother human-robot collaboration.

Training is paramount. Employees must understand how to safely interact with robots and automated systems. This goes beyond simple operation manuals. It involves hands-on training that covers safety procedures, emergency protocols, and collaborative robot (cobot) interaction guidelines. Regular refresher courses are crucial to maintain competency and awareness of evolving safety standards and best practices within the constantly evolving smart factory landscape. Investing in such comprehensive training programs is not simply a cost; it's a crucial investment in the safety and skill development of your workforce, directly impacting the overall efficiency and safety of your smart factory.

Beyond physical ergonomics, consider the cognitive aspects. The constant monitoring of automated systems can lead to mental fatigue. Implementing systems that provide clear and concise visual feedback, coupled with efficient alarm systems, can reduce cognitive overload. The design of the control interfaces, including the placement of screens and the clarity of information displayed, are critical for optimizing worker performance and minimizing cognitive strain. A well-designed control system should reduce decision fatigue and allow for quicker reaction times in critical situations, contributing to a safer and more productive work environment.

Don't forget about noise levels. Excessive noise in a factory environment can lead to hearing loss, stress, and reduced concentration. Implementing noise-reducing measures, such as sound-dampening materials and strategically placed noise barriers, can significantly improve the work environment. Regular noise level assessments are necessary to monitor compliance with safety regulations and to identify areas needing improvement. Reducing noise not only improves worker comfort and health but also enhances communication and collaboration within the factory, leading to a more efficient and safer work environment. Remember, a quiet factory is a productive factory.

Finally, regular audits and evaluations are critical. These aren't just boxes to tick; they're opportunities to proactively identify and address potential ergonomic and safety risks. Involving employees in these evaluations helps to foster a culture of safety and empowers them to raise concerns. Use these audits to refine existing procedures, implement new safety measures, and continually improve the overall workspace design. A continuous improvement mindset regarding safety is not optional; it's an integral part of a successful and sustainable smart factory. By prioritizing ergonomic and safety considerations, you're creating a workspace where technology and human talent work together seamlessly, safely, and productively.

5.2 Training and Skill Development

The transition to smart factories demands a workforce equipped for the challenges and opportunities ahead. It's not simply about replacing humans with machines; it's about forging a powerful partnership. This requires a fundamental shift in training and skill development, moving beyond traditional apprenticeships to encompass a broad spectrum of new technologies and collaborative work methodologies. Successful integration hinges on comprehensive training programs that equip employees with the necessary skills to operate, maintain, and collaborate with intelligent automation systems.

Upskilling existing staff is paramount. This involves tailored training modules focusing on specific technologies relevant to their roles, such as programming industrial robots, interpreting data from IoT sensors, or utilizing advanced analytics platforms. The curriculum should emphasize hands-on experience, simulating real-world scenarios to build confidence and proficiency. Furthermore, interactive simulations and virtual reality training can provide safe and cost-effective environments to learn complex procedures.

Beyond technical skills, soft skills are equally crucial. Collaborative robots, or cobots, demand a new level of human-machine interaction. Employees need training in effective collaboration with these intelligent machines, understanding their capabilities and limitations, and troubleshooting potential problems. This includes communication skills to effectively interact with the cobots and interpret their operational data. Problem-solving skills also take center stage, as employees must learn to identify and resolve issues arising from the complex interplay between human and robotic systems. Teamwork becomes even more critical, with multi-disciplinary teams collaborating to optimize production processes.

Developing a robust training infrastructure is essential. This goes beyond individual training modules, encompassing a holistic approach to learning and development. This may necessitate investing in state-of-the-art training facilities, equipped with the latest smart factory technologies. Partnering with educational institutions to develop specialized curriculum and certifications can also ensure a steady stream of skilled workers. Furthermore, continuous learning and development should be encouraged, providing employees with opportunities to upgrade their skills and stay abreast of the latest advancements in the field.

Effective training programs should also incorporate robust assessment methods, to ensure that employees are adequately prepared for their roles. This could include regular performance evaluations, both practical and theoretical tests, and ongoing feedback from supervisors. The assessment should not only focus on technical skills but also evaluate their ability to collaborate effectively with robots and solve problems in a dynamic environment. Continuous improvement should be integral, with regular reviews of the training program to assess its effectiveness and make necessary adjustments based on employee feedback and evolving technology. A truly effective program will not only train workers but also help develop their adaptability to a constantly changing technological landscape. This adaptability, a capacity for continuous learning, will be a defining characteristic of the successful smart factory workforce. Therefore, investing in training isn't just an expense; it's a strategic investment in a future-ready workforce, integral to the success of the smart factory.

The success of any smart factory initiative hinges on the skills and adaptability of its workforce. Ignoring the human element will invariably lead to underutilized technology and lost potential. A comprehensive, ongoing investment in training and development is not just a good practice, it is a necessity for successful integration and optimal utilization of advanced technologies in the modern manufacturing environment. The future-proofed workforce, capable of seamlessly integrating with advanced automation, will be the cornerstone of the smart factory's continued success and competitiveness. This requires a shift in mindset, from viewing training as a one-time cost to recognizing its importance as a continuous investment that yields long-term returns. The commitment to upskilling and reskilling employees should be viewed as a significant strategic advantage, driving productivity, innovation, and sustained growth.

5.3 Collaborative Robots (Cobots)

Cobots, or collaborative robots, represent a significant advancement in human-robot interaction within smart factories. Unlike traditional industrial robots operating in isolated cages, cobots are designed to work alongside human workers, sharing the same workspace. This collaborative approach offers numerous advantages, enhancing productivity and safety while reducing the need for extensive and potentially dangerous reconfigurations of the production line.

Their inherent safety features are paramount. Cobots are equipped with advanced sensors and software that enable them to detect and react to unexpected human presence or contact, automatically slowing down or stopping their movements. This reduces the risk of workplace accidents significantly, fostering a safer environment for everyone. This is particularly valuable in tasks involving repetitive movements or handling heavy objects where human error or strain can lead to injury.

Furthermore, the ease of programming and operation makes cobots an attractive option for businesses of all sizes. Many cobots utilize intuitive interfaces, requiring minimal specialized training to operate effectively. This reduces implementation costs and allows for faster deployment, maximizing return on investment. Their flexibility is also noteworthy; cobots can be reprogrammed and redeployed quickly, adapting to changing production needs without significant downtime or expense. This agility is vital in dynamic manufacturing environments where product variations or order volumes fluctuate frequently.

The implementation of cobots requires careful consideration of workspace design and ergonomics. Placing cobots strategically within the production line, taking into account human movement and reach, optimizes efficiency and reduces potential conflicts. Proper training of employees on safe collaborative practices is also crucial, ensuring workers understand how to interact effectively and safely with the robots. This training should include both theoretical knowledge about cobot operation and practical, hands-on experience working alongside the machines in controlled environments. Furthermore, regular safety assessments and audits are essential for maintaining a safe working environment.

One of the most significant advantages of cobots is their ability to enhance human capabilities. They can handle repetitive, strenuous, or hazardous tasks, freeing up human workers to focus on more complex and value-added activities requiring creativity, problem-solving, and decision-making. This improves overall productivity and allows workers to upskill and contribute to higher-level tasks, increasing job satisfaction and employee retention. This synergistic relationship between human and machine represents a significant shift in the approach to manufacturing. Think of a worker using a cobot to assist in assembling complex components – the robot handles precise placements while the human operator oversees the process and intervenes where needed. This combined efficiency far surpasses either human or robot working in isolation.

The choice of cobot model should be tailored to the specific needs of the manufacturing process. Factors such as payload capacity, reach, speed, and precision must be carefully evaluated to select the optimal cobot for each application. Consider the size and complexity of the components being handled; a cobot with a higher payload is necessary for heavier components while a robot with a longer reach is needed for larger workspaces. Speed and precision will depend on the level of accuracy required by the specific tasks. A detailed analysis of the production process is crucial to choose the appropriate cobot configuration, ensuring effective integration

into existing workflows.
Investing in cobots represents a long-term strategic decision for businesses looking to enhance their competitiveness. The initial investment might seem significant, but the return on investment (ROI) over time is substantial. Consider the reduced labor costs, increased productivity, improved product quality, and enhanced safety—all of these contribute to a significant positive impact on the bottom line. Furthermore, cobots can play a critical role in addressing labor shortages and attracting a new generation of skilled workers to the manufacturing industry, who are often more attracted to modern technology-rich environments. Businesses should carefully weigh the short-term expenses against the long-term benefits.
The integration of cobots into smart factories isn't just a technological upgrade; it's a fundamental shift in how manufacturing is perceived and conducted. It represents a move towards a more human-centered approach, where technology empowers human workers instead of replacing them. This integration fosters a more collaborative, efficient, and safer environment, improving overall productivity and competitiveness. As we move further into the era of Industry 4.0, the use of cobots will become increasingly prevalent, transforming the landscape of smart factories across various industries. Investing in this technology will not only enhance current operations but position businesses for success in the future.

6 Case Studies and Best Practices

Siemens' Amberg plant, a leader in smart manufacturing, showcases a remarkable transformation. Their implementation of highly automated systems, including sophisticated robotics and advanced analytics, drastically reduced lead times and improved product quality. This success, however, wasn't without its hurdles. Initial investment was substantial, requiring careful financial planning and stakeholder buy-in. Furthermore, integrating legacy systems with new technologies presented a significant technical challenge, necessitating a phased approach and expert IT support. The substantial training programs undertaken to upskill their workforce proved crucial for the long-term success of their smart factory initiatives. Their experience underlines the importance of a comprehensive strategy that addresses both technological and human capital aspects.

Consider the contrasting experience of a smaller manufacturer, a mid-sized automotive parts supplier in Detroit, Michigan, who faced different challenges. Their limited resources meant a more incremental approach was necessary. They started with a pilot project focusing on a single production line, implementing robotic welding and a real-time monitoring system. This allowed them to test and refine their processes before scaling up. The initial investment was less daunting, but their success hinged on a highly skilled team who were proactive in troubleshooting the inevitable technical difficulties and adapting their workflows to maximize the benefits of the new technology. The lean approach allowed for faster return on investment and minimized disruption to their operations. The result, an increase in efficiency and profitability of 15% within two years. This smaller-scale example highlights the adaptability of smart factory principles to varying organizational contexts.

Another compelling case study involves a pharmaceutical company in Switzerland that prioritized process optimization and quality control through digital twin technology. Using virtual models of their manufacturing processes, they could simulate different scenarios, identifying bottlenecks and optimizing workflows before implementing changes in the physical plant. This minimized downtime and avoided costly errors. This virtual approach also helped in demonstrating the value proposition of their smart factory upgrade to stakeholders. However, creating and maintaining accurate digital twins required significant expertise and considerable computing power, representing both a financial and technical investment. The return on investment, however, was substantial, with measurable improvements in product consistency and reduced waste, thus strengthening their competitive edge in a highly regulated market.

While these success stories are inspiring, many companies encounter unforeseen challenges during their smart factory journey. Data security breaches, for instance, are a significant concern. Companies need robust cybersecurity protocols and employee training to mitigate such risks. Unexpected equipment downtime can also disrupt production, highlighting the need for preventative maintenance and redundancy planning. Integrating new technologies into existing systems can be complex, often requiring extensive system integration efforts and potentially leading to unexpected delays. Furthermore, achieving the necessary expertise and skilled workforce to manage and maintain these advanced systems often involves significant investment in training and recruitment, adding complexity to the entire process. Proactive risk assessment, rigorous planning, and a flexible approach are therefore essential for successful smart factory implementation. Continuous improvement should be a cornerstone of the strategy.

The key takeaway from these examples isn't merely technological prowess, but a holistic approach. Successful smart factory implementation necessitates a multi-faceted strategy that encompasses careful planning, robust cybersecurity measures, comprehensive employee training, strategic partnerships, and a deep understanding of both the technological capabilities and the limitations of the chosen systems. These case studies underline the importance of adaptability and a willingness to learn from both successes and failures. Every organization's journey toward a smart factory is unique, shaped by its specific context, resources, and objectives. Therefore, a tailor-made strategy remains critical for sustained success in the dynamic landscape of smart manufacturing. The common thread among these successful organizations is a commitment to continuous improvement, innovation, and adaptability. By learning from the experiences of others, organizations can pave their own path toward more efficient, productive, and competitive manufacturing operations, unlocking the true potential of smart factory technologies.

6.1 Real-World Implementations

A German automotive manufacturer, renowned for its precision engineering, faced escalating production costs and shrinking profit margins. Their aging assembly lines, while reliable, lacked the flexibility and efficiency needed to compete in a rapidly changing market. Implementing a smart factory solution became not just a strategic advantage, but a necessity for survival. Their transformation began with a phased approach, starting with the integration of IoT sensors across their existing machinery. This allowed for real-time monitoring of production parameters, providing crucial data for process optimization. The initial investment was significant, requiring careful budget allocation and justification to stakeholders. However, the early returns were impressive, revealing bottlenecks and inefficiencies previously undetected. This data-driven approach led to immediate improvements in production flow, reducing downtime and waste.
The next phase involved the deployment of collaborative robots, or cobots, alongside human workers. This wasn't a simple replacement of human labor, but a carefully planned integration designed to leverage the strengths of both human ingenuity and robotic precision. Extensive training programs were implemented, ensuring that workers felt empowered, not threatened, by the introduction of automation. This proactive approach fostered a culture of acceptance and collaboration, mitigating po-

tential workforce anxieties often associated with technological advancements in the manufacturing sector. The result was a more efficient and safer work environment, where humans handled tasks requiring complex decision-making and problem-solving while robots handled repetitive, high-precision tasks.

Another compelling example comes from a food processing company in the United States. Their challenge revolved around maintaining consistent product quality while meeting fluctuating demand. Traditional quality control measures were proving insufficient, leading to significant product waste and customer complaints. Their solution involved implementing advanced machine learning algorithms to analyze sensor data from their processing lines. This allowed for predictive maintenance, preventing equipment failures before they occurred, and real-time adjustments to processing parameters to ensure consistent product quality, regardless of demand variations. The deployment of machine vision systems further enhanced quality control, automatically identifying and rejecting substandard products. This proactive approach minimized waste, reduced costs, and significantly improved customer satisfaction. The implementation also required rigorous data security measures to protect sensitive production data.

These case studies showcase the diverse applications of smart factory technologies. They demonstrate that successful implementation requires a holistic approach, encompassing technological integration, process optimization, workforce training, and robust cybersecurity measures. The benefits are clear: increased efficiency, reduced costs, improved quality, enhanced responsiveness to market demands, and a safer, more engaging work environment for employees. The critical factor, however, is careful planning, phased implementation, and a commitment to continuous improvement. It's not simply about replacing old machinery with new; it's about fundamentally transforming the way production is managed and optimized. Choosing the right technology for specific needs, rather than adopting a one-size-fits-all approach, is paramount. Similarly, understanding the human element – addressing worker concerns, providing adequate training, and fostering a culture of collaboration – is crucial for a successful and sustainable smart factory transformation. Ignoring these

aspects could potentially negate many of the expected benefits.
Furthermore, successful implementations often rely heavily on strong partnerships. Collaboration with technology providers, system integrators, and even competitors can accelerate the learning curve and mitigate potential risks. These partnerships often provide access to specialized expertise, cutting-edge technologies, and best practices, streamlining the implementation process and minimizing disruptions to ongoing operations. Sharing knowledge and experiences across industries can also foster innovation and drive the adoption of new technologies. Building a robust network of support is therefore crucial for navigating the complexities of smart factory transformation, providing access to knowledge and resources that can ensure long-term success. Continuous evaluation and adaptation are also vital to sustain the improvements made during the implementation. Smart factories are dynamic environments; the need to adapt and improve should never be neglected. The learning process should never stop.

6.2 Lessons Learned and Challenges

Implementing smart factory systems presents a unique set of challenges. Many companies, enthusiastic about the promise of increased efficiency and reduced costs, often underestimate the complexities involved in integrating diverse technologies, adapting existing workflows, and managing the transition. Successful implementation hinges on a comprehensive understanding of these hurdles, necessitating careful planning, robust risk assessment, and a commitment to continuous improvement. Ignoring these aspects can lead to delays, budget overruns, and even project failure. This isn't just about technology; it's about people, processes, and a cultural shift within the organization.
Data integration, a cornerstone of a successful smart factory, proves surprisingly difficult. Harmonizing data from various sources—machines, sensors, ERP systems—requires meticulous planning and the development of standardized data formats. Inconsistent data formats hinder real-time analysis and decision-making, limiting the benefits of smart factory technologies. This often necessitates significant investment in data management infrastructure and skilled personnel capable of handling the com-

plexities of big data analytics. Furthermore, integrating legacy systems, sometimes decades old, into a modern, interconnected network poses its own set of unique technological and logistical obstacles.

The initial investment cost represents a substantial barrier for many businesses, particularly smaller or mid-sized enterprises. The expense extends beyond the purchase of new equipment; it encompasses software licenses, integration services, employee training, and ongoing maintenance. A lack of sufficient funding, or improper financial planning, can cripple a smart factory initiative before it even gets off the ground. Carefully considering return on investment (ROI) and developing a phased implementation plan—starting with smaller, more manageable projects—can help mitigate this risk, making the transition more financially feasible.

Cybersecurity is a paramount concern, and one that is frequently overlooked during the initial stages of planning. The interconnected nature of smart factories creates an extensive attack surface, making them vulnerable to various cyber threats. From malware infections to data breaches, the consequences of a successful cyberattack can be devastating, resulting in production downtime, financial losses, and reputational damage. A proactive cybersecurity strategy, incorporating robust security measures and protocols, is crucial throughout the entire lifecycle of a smart factory project, from design and implementation to ongoing maintenance. This requires specialized expertise and ongoing investment.

Change management is another often-underestimated challenge. Introducing new technologies and processes inevitably disrupts the status quo, affecting employees at all levels. Resistance to change is a natural human response, and without careful planning, this resistance can undermine the entire project. Effective communication, thorough training, and a demonstrable commitment to employee well-being are vital to mitigating this resistance. The successful integration of smart factory systems depends not only on the seamless functioning of technology but also on the effective management of the human element, including worker upskilling and retraining programs. Addressing these concerns head-on helps ensure a smoother, more successful transition.

Maintaining the complex systems of a smart factory is an ongoing process that requires specialized expertise and significant resources. The intricate

interplay of various technologies and the constant influx of data necessitates a skilled team of technicians, engineers, and data scientists capable of ensuring optimal system performance and addressing any arising issues promptly. This requires a significant investment in training and development, ensuring that the workforce possesses the necessary skills to maintain and optimize the factory's complex systems. Furthermore, the need for continuous software updates, system upgrades, and proactive maintenance adds to the operational overhead of running a smart factory.
Finally, the evolving nature of technology necessitates a commitment to continuous improvement and adaptation. Smart factory technology is constantly evolving, with new technologies and techniques emerging regularly. Successful adoption requires a long-term perspective, a commitment to ongoing innovation, and the flexibility to adapt to new advancements as they become available. This can involve retraining employees, upgrading systems, and constantly evaluating the efficiency and effectiveness of implemented solutions. This ongoing process ensures that the smart factory remains competitive and up-to-date with industry best practices. The journey towards a truly smart factory is a continuous process of learning, adaptation, and optimization.

7 Future Trends in Smart Factories

Additive manufacturing, or 3D printing, will revolutionize production processes. Imagine, instead of relying on large, expensive molds, creating customized parts on demand. This drastically reduces lead times, minimizes waste, and opens doors to highly personalized products. Furthermore, the integration of AI-driven design optimization tools will further enhance the capabilities of 3D printing, allowing for the creation of lighter, stronger, and more efficient components tailored to specific needs. This capability will reshape entire industries, from aerospace to medical devices, making bespoke solutions economically viable. We're already seeing initial advancements in materials science that will expand the scope of 3D-printed parts, allowing for the production of components with superior properties, exceeding the capabilities of traditional manufacturing methods. Imagine the possibilities.

The rise of autonomous mobile robots (AMRs) will transform material handling and logistics within smart factories. Unlike traditional automated guided vehicles (AGVs), which follow fixed paths, AMRs utilize advanced sensors and AI to navigate dynamically, adapting to changing environments and avoiding obstacles. This dynamic adaptation enhances efficiency and flexibility, increasing responsiveness to changing production needs. Their use extends beyond simple material transport; AMRs can be equipped to perform a variety of tasks, such as inventory management, quality control inspections, and even assisting human workers with physically demanding jobs. This results in a significant improvement in workflow efficiency and the reduction of labor-intensive tasks. They're capable of operating 24/7, significantly boosting productivity, and can be integrated seamlessly with other smart factory systems, creating a truly interconnected and responsive manufacturing environment.

Sustainability will become a paramount concern, pushing the adoption of green manufacturing practices. This will involve the use of renewable energy sources to power factories, reducing carbon footprints and dependence on fossil fuels. Recycling and waste reduction initiatives will be implemented rigorously, minimizing environmental impact. Smart factories will be designed to optimize energy consumption, implementing real-time monitoring and control systems to identify and address inefficiencies. The use of sustainable materials will also become standard practice, moving away from environmentally damaging resources and promoting the adoption of bio-based materials and recycled components. Moreover, the entire production process will be scrutinized for environmental implications, driving innovation in cleaner manufacturing methods and reducing overall pollution. This move towards sustainable manufacturing will not just be an ethical imperative, but a competitive advantage, attracting environmentally conscious consumers.

The future workforce will need to adapt to the changing landscape. Reskilling and upskilling initiatives will become vital, ensuring that the workforce possesses the necessary digital literacy and technical skills to operate and maintain sophisticated smart factory systems. This includes expertise in areas such as data analytics, robotics programming, cybersecurity, and AI. Moreover, new roles will emerge, requiring individuals with specialized knowledge in areas like predictive maintenance, digital twin modeling, and systems integration. Educational institutions will need to adapt their curricula to meet the evolving demands of the smart factory environment, bridging the gap between academic training and industry requirements. Life-long learning will become crucial for individuals to remain relevant and competitive in the smart manufacturing workforce.

Global collaboration will drive innovation in smart factories. This will involve sharing best practices, technologies, and data across different manufacturing facilities and companies. International partnerships will become increasingly important, enabling the development of standardized protocols and interoperable systems. Data exchange platforms will facilitate seamless communication and information flow, allowing businesses to collaborate effectively and learn from each other's experiences. This globalized approach fosters innovation, accelerates technological adoption, and ultimately, benefits the entire manufacturing industry. Ultimately, this shared knowledge will lead to the development of more efficient, sustainable, and adaptable smart factories globally.

7.1 Emerging Technologies

Additive manufacturing, or 3D printing, is rapidly transforming the manufacturing landscape. No longer limited to prototyping, this technology now allows for the production of complex, customized parts directly from digital designs, eliminating the need for traditional tooling and significantly reducing lead times. This offers unparalleled flexibility, enabling on-demand production and reducing inventory costs, a clear advantage over conventional methods. Consider the potential for personalized medicine, where implants or prosthetics can be precisely tailored to individual patient needs, produced locally, and on demand. This level of customization was previously unimaginable, and now it's becoming a tangible reality.
Blockchain technology, while often associated with cryptocurrencies, holds immense potential for enhancing supply chain transparency and security in smart factories. By creating a secure, immutable record of every stage of the production process, from raw material sourcing to final product delivery, blockchain eliminates the possibility of counterfeiting and improves traceability. This increased visibility allows for rapid identification and resolution of supply chain disruptions, minimizing downtime and improving overall efficiency. Imagine the impact of real-time tracking of components, ensuring quality and preventing the use of substandard materials. This enhances not only product quality but also consumer trust and brand reputation.
Advanced robotics are moving beyond simple repetitive tasks. Collaborative robots, or cobots, are designed to work safely alongside human workers, enhancing productivity and reducing the risk of workplace injuries. These robots are equipped with advanced sensors and AI-powered

capabilities, allowing them to adapt to changing environments and collaborate effectively with humans on complex tasks. The development of more sophisticated AI algorithms further improves their capabilities, allowing for increased autonomy and improved efficiency in various industrial settings. This is crucial for addressing the skills gap in the manufacturing industry and improving overall safety. The enhanced precision and repeatability of robotic systems also contribute to higher quality products.

The integration of augmented reality (AR) and virtual reality (VR) technologies offers new possibilities for training, maintenance, and remote collaboration in smart factories. AR overlays digital information onto the real world, providing workers with real-time guidance during complex tasks or repairs, significantly reducing training times and error rates. VR, meanwhile, allows for simulated training environments, enabling workers to practice complex procedures in a safe, controlled setting before attempting them in the real world. This minimizes risks associated with real-world errors and improves the overall efficiency of training and maintenance processes, a truly remarkable advancement. Remote collaboration via AR and VR also allows for expertise to be readily accessible across geographic locations.

Quantum computing, still in its nascent stages, promises to revolutionize various aspects of smart factory operations. Its immense computational power could optimize complex processes such as supply chain management, predictive maintenance, and materials science, leading to breakthroughs in efficiency, cost reduction, and product innovation. While still years away from widespread adoption, the potential of quantum computing to solve currently intractable problems is undeniable, promising a future where smart factories operate with unparalleled efficiency and precision. The advancements in this field are shaping the very foundation of future manufacturing practices. This technology will enable the solving of problems previously deemed unsolvable.

The convergence of these emerging technologies is creating a synergistic effect, further enhancing the capabilities of smart factories. The integration of additive manufacturing, blockchain, advanced robotics, AR/VR, and quantum computing will lead to unprecedented levels of automation, efficiency, and customization in manufacturing, transforming the industry in profound ways. The future of manufacturing is dynamic and evolving, continually seeking new ways to achieve perfection and address emerging challenges. The continuous adoption of new technologies is fundamental to this evolution, driving improvements and breakthroughs.

7.2 Sustainability and Green Manufacturing

Adopting sustainable practices isn't merely an ethical choice; it's a strategic imperative for smart factories. Resource efficiency translates directly to cost savings, reduced environmental impact enhancing brand reputation, and attracting environmentally conscious customers. Companies embracing sustainability often find themselves ahead of the curve, setting industry standards and gaining a competitive edge. These advantages are significant and long-lasting, affecting the entire lifecycle of a product.

The transition to green manufacturing requires a multifaceted approach. It begins with selecting eco-friendly materials, prioritizing those sourced responsibly and with minimal environmental impact. This involves careful consideration of the entire supply chain, from raw material extraction to final product disposal. Companies should actively seek suppliers committed to sustainability and transparency, fostering collaborative relationships based on shared environmental goals. This collaborative effort extends beyond the immediate supply chain, engaging with partners and stakeholders across the entire value network.

Energy efficiency is another crucial component. Smart factories, with their interconnected systems, offer unique opportunities for optimization. Real-time monitoring of energy consumption, coupled with data analytics, allows for the identification of inefficiencies and the implementation of targeted interventions. This might involve optimizing production schedules to minimize energy peaks, implementing energy-efficient equipment, or exploring renewable energy sources such as solar or wind power. Adopting smart energy management systems is vital for long-term cost savings and a reduced carbon footprint.

Waste reduction is paramount. Smart manufacturing technologies offer significant potential for minimizing waste generation throughout the production process. Data-driven insights enable proactive identification and prevention of defects, reducing the amount of scrap materials and improving overall product yield. Implementing lean manufacturing principles, along with advanced quality control systems, further minimizes waste and optimizes resource utilization. Recycling and repurposing of materials should also be prioritized, closing the loop and contributing to a circular economy.

Water conservation is another critical aspect of green manufacturing. Smart factories can employ technologies to monitor and optimize water usage, identifying leaks and inefficiencies. Implementing water-saving processes, such as closed-loop water systems, minimizes water consumption and protects water resources. Careful consideration of water usage in each manufacturing stage is crucial for efficient resource management and environmental protection.

Beyond these operational considerations, businesses should actively invest in carbon footprint reduction strategies. This could involve implementing carbon capture technologies, investing in renewable energy sources, or offsetting emissions through verified carbon offsetting programs. Regular carbon footprint assessments, coupled with transparent reporting, provide accountability and demonstrate commitment to sustainable practices. Such assessments should extend beyond the immediate factory operations, incorporating the entire supply chain for a comprehensive environmental impact assessment.

Green manufacturing extends beyond simple operational changes. It requires a fundamental shift in organizational culture. This involves integrating sustainability into all aspects of business operations, from procurement and design to production and disposal. Employee training and engagement are essential to build a shared understanding and commitment to sustainable practices. This can involve creating a sustainability team, incorporating sustainability metrics into performance evaluations, and actively promoting environmental awareness within the workforce. Encouraging employee participation in green initiatives fosters a culture of responsibility and drives innovation.

Furthermore, transparency and communication are crucial. Companies should openly communicate their sustainability efforts to stakeholders, demonstrating accountability and building trust. This could involve publishing sustainability reports, engaging with environmental organizations, and participating in industry initiatives aimed at promoting sustainable manufacturing practices. Open and honest communication builds confidence and reinforces the company's commitment to a greener future.

Finally, a long-term perspective is essential for successful green manufacturing. It is not a one-time project, but an ongoing commitment that requires continuous improvement and adaptation. Regular reviews of sustainability initiatives, coupled with data-driven analysis, allow for adjustments and optimization. Investing in research and development, exploring new technologies, and proactively anticipating future environmental regulations are critical for maintaining a competitive edge in the ever-evolving landscape of sustainable manufacturing. The ultimate goal is to build a sustainable and resilient business model that contributes to a greener future for all.

7.3 The Future Workforce

The shift to smart factories necessitates a fundamental change in the workforce. We're not just talking about adding new technologies; we're talking about reshaping roles, responsibilities, and even the very skills needed to thrive in this new manufacturing landscape. The demand for traditional manufacturing roles, those focused solely on repetitive tasks, will inevitably decrease. This isn't a threat, but an opportunity to elevate the human element within the factory.

Instead of focusing on physical, manual labor, the future workforce will heavily emphasize higher-level skills. Think about data analysis, programming, and system maintenance. Workers will need to understand the complex interplay of interconnected systems, possessing the analytical skills to interpret vast amounts of data generated by the smart factory's numerous sensors and machines. Troubleshooting complex automation systems and identifying potential malfunctions will become paramount. Proactive problem-solving and predictive maintenance will be key skills, preventing costly downtime and improving overall efficiency.

This necessitates a significant investment in education and training. Existing employees will require upskilling programs to equip them with the necessary digital literacy and technical expertise. This isn't a one-time investment; it's an ongoing process of continuous learning and adaptation. Collaborations between educational institutions, industry leaders, and government bodies will be crucial in developing relevant curricula and providing accessible training opportunities. The curriculum should focus not just on technical skills, but also on problem-solving, critical thinking, and teamwork—all essential for effective human-robot collaboration.

Furthermore, the future workforce will need to be adaptable and versatile. The rapid pace of technological advancement in smart factories means that new technologies and processes will constantly emerge. Workers will need to embrace lifelong learning, constantly acquiring new skills and knowledge to remain relevant and competitive. This adaptability extends beyond technical skills; it encompasses a broader capacity for change and innovation. The ability to learn quickly, adjust to new situations, and embrace new challenges will be highly valued.

Beyond technical skills, the human element remains crucial. Soft skills such as communication, teamwork, and problem-solving will remain in high demand. These skills are critical for effective human-robot collaboration, ensuring seamless integration and maximizing productivity. The ability to collaborate effectively with robots and other automated systems, understanding their capabilities and limitations, will be a defining characteristic of the future workforce. Moreover, strong communication skills will be essential for reporting data insights and coordinating with colleagues across different departments.

The human-centered design of smart factories ensures that the transition is not only technologically advanced but also employee-focused. Ergonomics and workplace safety must remain paramount, ensuring the physical and mental well-being of the workforce. The design of the smart factory itself will need to adapt to enhance human-robot interaction, creating collaborative spaces that foster a sense of shared purpose. This includes creating accessible interfaces and user-friendly systems for employees to easily interact with the technology and understand the data being generated. Human factors engineering will play a critical role in this human-centric approach to smart factory design.

Ultimately, the future workforce in smart factories will be a blend of high-tech expertise and strong human skills. It will be characterized by adaptability, continuous learning, and a focus on collaboration. Investing in workforce development and creating a human-centered work environment will be essential for the successful implementation and operation of smart factories, ensuring both productivity and employee well-being. The smart factory of the future is not just about intelligent machines; it's about intelligent humans working alongside them, creating a truly symbiotic and productive environment. The success of this transition hinges on the preparedness and adaptability of the workforce, making it a vital area of focus for businesses and policymakers alike. By embracing a proactive and holistic approach to workforce development, organizations can ensure a smooth and successful transition to the era of smart manufacturing, maximizing the benefits of intelligent automation while preserving the invaluable contribution of the human element.

7.4 Global Collaboration

The interconnected nature of global supply chains necessitates a collaborative approach to smart factory development. Successfully implementing and scaling smart factory technologies requires a global network of expertise, sharing best practices and overcoming common challenges. This isn't just about exchanging data; it's about fostering a truly collaborative ecosystem. Imagine, for instance, a Japanese manufacturer sharing its expertise in precision robotics with a German counterpart specializing in advanced sensor technology, leading to the development of a superior, more efficient robotic arm. This kind of cross-pollination of ideas and resources accelerates innovation far beyond what any single entity could achieve alone.
Effective global collaboration demands a clear strategy. Companies need to identify partners strategically, selecting organizations with complementary skills and resources. This isn't about simply finding the cheapest option; it's about finding the right fit, ensuring synergy and minimizing potential conflicts. A thorough due diligence process is critical, vetting potential partners to assess their reliability, security protocols, and alignment with ethical manufacturing principles. Establishing clear communication channels and shared goals is crucial for avoiding misunderstandings and

delays. Open communication platforms, such as secure online collaboration tools, allow for seamless information sharing, regardless of geographical location. Regular virtual meetings, utilizing video conferencing technology, allow for real-time discussion and problem-solving.

Building trust is paramount in any successful partnership, especially across international borders. This requires transparency in operations and a shared commitment to ethical practices. Openly discussing challenges, successes, and failures fosters a culture of mutual support and understanding. Sharing best practices and lessons learned speeds up the learning curve for all involved, preventing costly mistakes and allowing for more efficient development. Strong legal agreements and intellectual property protection frameworks are crucial to safeguarding the interests of all participants. These agreements must outline responsibilities, share ownership, and address potential conflicts.

Globalization presents challenges, however. Different regulatory landscapes across countries can complicate things. Data privacy regulations, for instance, vary widely, necessitating careful planning to ensure compliance across all participating jurisdictions. Differences in labor laws and cultural norms also demand sensitivity and adaptability. Building a diverse and inclusive team, drawing on expertise from different regions, helps to navigate these complex issues more effectively. By respecting cultural differences and promoting inclusive working environments, the collaborative project becomes more robust and adaptable to a wider range of situations.

Overcoming language barriers can also be a significant hurdle. Implementing multilingual communication platforms and employing skilled translators ensures that information is accurately conveyed, avoiding misunderstandings and potential conflicts. Investing in training programs to enhance language skills for team members enhances project efficiency and understanding. The adoption of a common project management methodology, with clear roles and responsibilities, ensures that everyone works towards the same goals, despite language differences. Successful global collaboration in smart factory development demands a proactive approach to managing these potential obstacles. Open communication and mutual respect for diverse perspectives is key.

Beyond the practical considerations, cultivating a collaborative spirit amongst partners is crucial. This involves building relationships and fostering mutual respect. Regular social interaction, even through virtual channels, helps strengthen these relationships, leading to a more effective and productive collaboration. Joint training programs, workshops, and knowledge-sharing events contribute to a strong sense of community among the participants, which fosters innovation and improves problem-solving. Regular feedback sessions and performance evaluations help build trust and accountability, ensuring that the collaborative venture remains focused on its primary goals. In essence, global collaboration isn't just about technology; it's about building enduring relationships and harnessing the collective intelligence of a global community to create a more efficient, sustainable, and innovative future for smart factories.

7.5 Ethical Considerations

The rapid advancement of smart factory technologies presents exciting opportunities, but also necessitates careful consideration of ethical implications. We must proactively address potential biases embedded within algorithms, ensuring fairness and avoiding discriminatory outcomes in automated decision-making processes. For instance, AI-driven systems used for hiring or performance evaluation must be rigorously tested for bias, preventing the perpetuation of existing inequalities. Transparency in algorithmic processes is crucial, allowing for accountability and enabling stakeholders to understand how decisions are reached. This includes providing clear explanations for automated actions and allowing for human oversight when necessary, ensuring that the human element isn't entirely removed.

Data privacy is paramount. The vast amounts of data collected in smart factories must be handled responsibly, adhering to strict privacy regulations and ensuring the security and confidentiality of sensitive employee information. Robust data anonymization techniques should be implemented wherever possible, minimizing the risks associated with data breaches and misuse. Furthermore, clear and transparent data usage policies should be established, communicating how data is collected, used, and protected,

thereby building trust with employees and other stakeholders. Open communication about data practices is key to maintaining ethical operations.
Job displacement due to automation is a legitimate concern. While smart factories can increase overall efficiency and productivity, the transition to automated systems should be managed responsibly, minimizing job losses and providing opportunities for workforce reskilling and upskilling. Proactive measures, such as offering training programs and supporting employee transitions to new roles, are essential for mitigating the social and economic consequences of automation. Collaboration between businesses, educational institutions, and government agencies can help create a smooth transition, preparing the workforce for the changing demands of the smart factory environment.
The responsible use of resources is crucial. The shift towards smart factories should prioritize sustainable practices and minimize environmental impact. Optimizing energy consumption through smart energy management systems is vital. Choosing eco-friendly materials and reducing waste through efficient production processes contributes to a more sustainable manufacturing model. Investing in renewable energy sources and implementing circular economy principles can further minimize the environmental footprint of smart factories, aligning with broader sustainability goals. This proactive approach demonstrates ethical responsibility towards the environment.
Accountability for malfunctions and errors is a significant ethical challenge. The complex interconnectedness of smart factory systems means that determining responsibility for malfunctions or errors can be challenging. Clear lines of accountability should be established, encompassing all stakeholders involved in the design, implementation, and operation of the system. This requires the development of robust monitoring and auditing mechanisms to track system performance and identify potential failures promptly, enabling swift responses and effective rectification strategies. Furthermore, establishing clear procedures for investigating and addressing system failures is essential for ensuring accountability and building trust.
The ethical considerations discussed here are not merely abstract concepts; they are crucial for the responsible development and implementation of

smart factory technologies. Ignoring these ethical considerations not only jeopardizes the workforce and the environment but also damages the reputation and long-term viability of businesses. By proactively addressing these challenges, we can ensure that the benefits of smart factories are realized while upholding ethical principles and promoting a responsible and sustainable future. Only through a holistic approach, integrating technological advancements with ethical awareness, can we harness the full potential of smart factories without compromising our values or the well-being of society. The successful integration of ethical considerations into the design and operation of smart factories will be a defining factor in determining their long-term success and societal acceptance. It's not just about efficiency and productivity; it's about building a future where technology serves humanity ethically and responsibly. The ongoing dialogue and collaboration between technologists, policymakers, and society are vital for navigating the ethical complexities of this transformative technology.

www.ingramcontent.com/pod-product-compliance
Ingram Content Group UK Ltd.
Pitfield, Milton Keynes, MK11 3LW, UK
UKHW020155290926
15471UKWH00005B/888

9 798349 408229